W9-AGJ-204

PRINCETON ■ LONDON

What's in the book?

How Earth began 4
Inside Earth 6
Continent 8
Earthquake 10
Volcano 12
Rock 14
Erosion 16
Ocean 18
Desert 20
Polar lands 22
Grasslands 24
Forest 26
River 28
Mountain 30
Plant life 32
Animal life 34
Weather 36
Climate and season 38
Earth and its future 40
Where on Earth? 42
Amazing facts 44
Index 46
Troubleshooting tips 48

What's on the disk?

Eleven detailed maps look at the extraordinary variety of natural features found on Earth. Each interactive map leads to a related game or activity (see right). The disk's extensive Glossary explains the meaning of **bold** words in the book. It also provides key information on Earth-related words that are used in the disk's activities and topics.

Explore the main screen to find interactive maps, games, and activities.

Installing the Earth CD-ROM

See page 48 for troubleshooting tips, system requirements, and helpline details.

Windows 95 or 98

The Earth program should start automatically when you put the CD into your CD-ROM drive. If it does not, follow these instructions.

1. Put the CD into the CD drive.
2. Double-click on My Computer.
3. Double-click on the CD drive icon.
4. Double-click on the EARTH icon.

Windows 3.1 or 3.11

1. Put the CD into the CD Drive.
2. Open File Manager.
3. Double-click on the CD drive icon.
4. Double-click on the EARTH icon.

Macintosh

1. Put the CD into the CD drive.
2. Double-click on the EARTH FOR MAC icon.

Power Macintosh

1. Put the CD into the CD drive.
2. Double-click on the EARTH FOR POWER MAC icon.

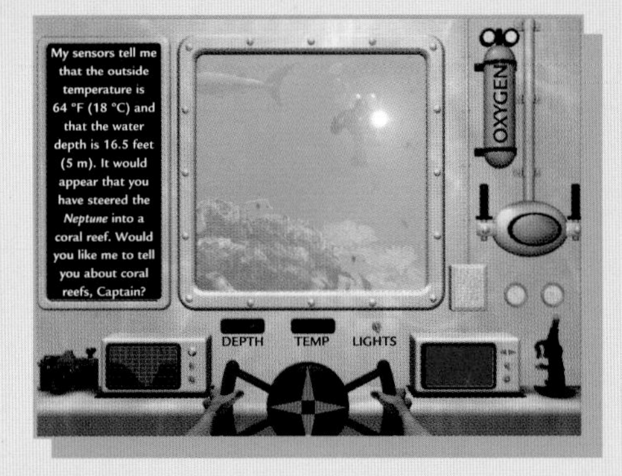

Underwater Mission

Steer a submersible through the ocean and photograph some of the mysteries of the deep. Use the submersible's onboard computer for clues to help you.

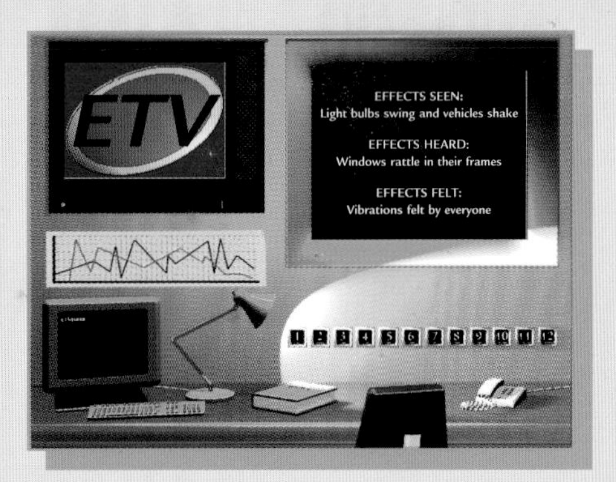

Earthquake Institute

Visit the Earthquake Institute and help them study tremors from around the world. Then witness the earthquake in a TV report that you helped create!

Blow your top

Will Mount Magmacone erupt spectacularly, showering the land with molten lava, or will it remain dormant? Only you can create the right conditions for an eruption.

Course to the source

Travel the length of the mighty Amazon River in this fast-flowing tropical adventure. You'll need expert map-reading skills and a thorough knowledge of rivers to make it to the source!

Jigsaw

Pick up the pieces and put the world back together with these interactive jigsaws. Then click away to find out more about the world's tectonic plates and climate zones.

Higher or lower

This TV game show is a card game with a difference! We've turned the world's mountains, rivers, and minerals into playing cards — it's up to you to decide whether the next card will be higher or lower.

The Toolbar

The toolbar appears whenever you move your cursor to the right side of the screen.

Click here to return to the main screen.

Click here to see your Geopass.

Click here to visit the museum shop.

Click here to use the glossary.

Click here to make notes.

Click here for help.

Click here to quit.

Under the weather

Steer your character through the layers of the atmosphere in this weather game. But don't let your judgment become clouded — you may get blown off course!

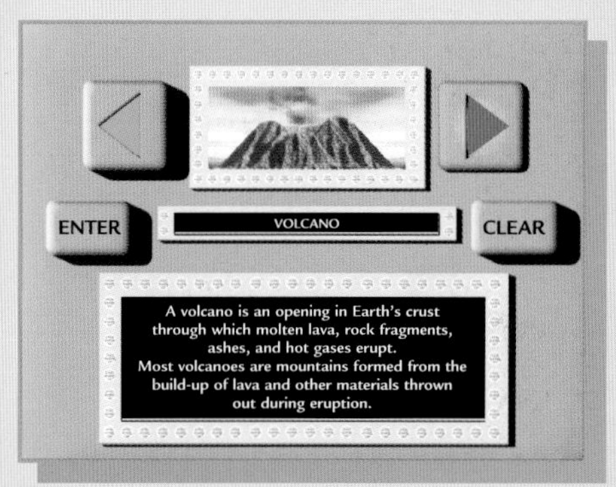

Glossary

All the Earth-related words you need to know are collected in the disk's Glossary. Just type in the word you don't understand and the interactive Glossary will explain it for you.

How Earth began

Earth is the name of the **planet** on which we live. It is a huge ball of rock spinning through space. No one is sure when Earth began, but some scientists think that, between 10 and 20 **billion** years ago, there was a huge explosion, called the **big bang**. Gradually, over millions of years, stars and planets started to form. One of these planets was Earth.

▼ Some scientists believe that, Earth formed in this way, about 4.5 billion years ago.

2 Over millions of years, grains of dust and ice in the clouds stuck together to make bigger and bigger pieces.

1 Earth may have started as clouds of **gas** and dust swirling around the sun.

▼ How life began

Life on Earth probably started in the oceans between 3 and 4 billion years ago. At first, there were only plants and simple living creatures called bacteria. Later, animals without backbones appeared, then, animals with backbones. Gradually, life moved from the ocean to the land.

Over 1,000 million years ago, animals with soft bodies, such as jellyfish and worms, developed in the sea.

By 500 million years ago, fish had appeared. Then, 370 million years ago, fishlike creatures that breathed air moved from the ocean onto the land.

3 Slowly, an enormous round ball of hot, fiery rock formed. Earth was surrounded by a layer of gases.

4 Eventually, the surface cooled and formed a hard, rocky **crust**. Drops of water in the air made clouds. Rain fell into dips on Earth's surface to make oceans.

Around 230 million years ago, dinosaurs roamed Earth. Other **reptiles**, such as crocodiles and turtles, also developed around this time.

Large **mammals** and birds appeared on Earth about 140 million years ago. Together, they took the place of the dinosaurs.

Most scientists believe that the first humans lived about 2 million years ago. These people made the first tools. Modern looking humans have been around for only about 100,000 years.

Go to How Earth began page 4, Rock page 14, Volcano page 12

Inside Earth

Planet Earth is a ball of rock and metal, made up of four layers. The thin, cool, outside layer is called the **crust**. People live on the surface of the crust. The three layers inside Earth are thicker and much hotter than the crust. No one knows exactly what Earth is like deep inside, but we do know that the hottest part is in the middle.

land and ocean
Most of the crust is covered by ocean, but other parts are dry land with soil on top.

Earth's layers
This picture shows the layers that make up Earth. The thickest layer lies just under the crust and is called the **mantle**. Below the mantle, there is an outer and an inner core.

Earth's crust
The thin crust is made of solid rock. It forms a skin, like the skin of an apple, around the planet.

mantle
This is a thick layer of rock. Rock close to the crust is melted and flows up and down, inside Earth.

outer core
Outside the solid inner core, there is a layer of liquid metals that is fiercely hot.

inner core
Earth's center is a ball of solid metal that may be as hot as 13,000 °F (7,000 °C).

Rocks inside Earth
Melted rock that flows in Earth's mantle is called **magma**. Sometimes, magma pushes its way up through a crack in Earth's crust, called a **volcano**. Then, it flows along the ground like a hot, red river. When magma reaches the surface, it is called **lava**.

Soil on the surface
Soil is made up of tiny pieces of rock from Earth's crust. It also contains plant and animal remains that make the soil rich and help plants to grow.

tunnels
Worm burrows let air and water into the soil.

roots
Plant roots help to hold the soil together.

bedrock
Beneath the soil lies the solid rock of Earth's crust. This is called bedrock.

Mining Earth
Deep within Earth, there are fuels such as **oil** and **gas**. They formed millions of years ago from rotted plants and animals. Today, miners drill down to 1,640 feet (5,000 m) to find oil and gas trapped between layers of rock. Then the fuel is used to make electricity in power stations.

drilling tower

layers of rock

drill pipe

gas

oil

Quarries
At the surface, miners dig large pits called quarries. Rocks are broken up by huge machines and used for building. Metals are taken from the rocks to make machinery.

Go to Earthquake page 10, Mountain page 30, Ocean page 18, Where on Earth? page 42

Continent

A continent is a huge area of land. There are seven continents on Earth. Each continent sits on top of a giant piece of crust called a plate. There are 19 large plates and a few smaller ones. Every year, the plates move a few inches. Over millions of years, these movements make the oceans and continents shift and change shape.

African Plate
This plate is called the African Plate. It includes the part of the Earth's crust under the continent of Africa and some crust under the ocean.

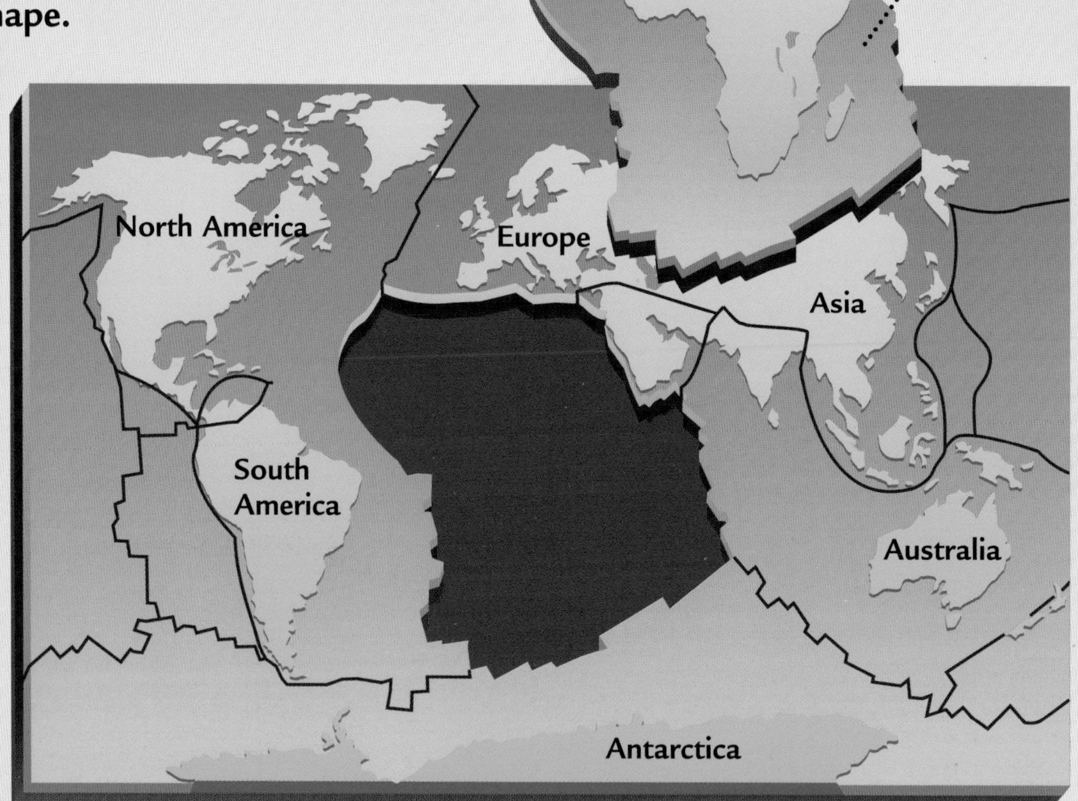

▶ This map shows the seven continents and some of the plates that lie beneath them.

Plates
Earth's plates fit together like a jigsaw puzzle. Plates float on top of Earth's **mantle**, like rafts on the sea. Some plates have ocean on top, others have a continent, or part of a continent, as well as ocean.

Plates pushing together
When two of Earth's plates crash into each other, the crust at the edge of the plates crumples and folds. Over millions of years, the crust is gradually pushed upward into high mountain peaks. This is how the Himalaya was formed. The Himalaya is the world's highest mountain range. They are still growing taller, because the plates are still pushing together.

Plates pulling apart

On some parts of the Earth's crust, huge plates pull apart. This happens mostly under the ocean, but sometimes plates pull away from each other under the land. Iceland is a country that sits on two plates that are pulling apart. This has made a long, deep crack in the land, as shown in the photograph. Eventually, Iceland will split in two along this crack, but this will not happen for millions of years.

Sliding plates

On other parts of Earth's crust, plates slide past each other and move slowly in opposite directions. This can make a deep crack in the land. This photograph shows the San Andreas **Fault** in California. Two of Earth's sliding plates meet here and earthquakes often happen.

Continents in the past

Scientists think that about 200 million years ago, all Earth's land fitted together in one big piece. This was a supercontinent called Pangaea. As the plates moved, they slowly pulled Pangaea apart, making new, smaller continents.

Pangaea 200 million years ago

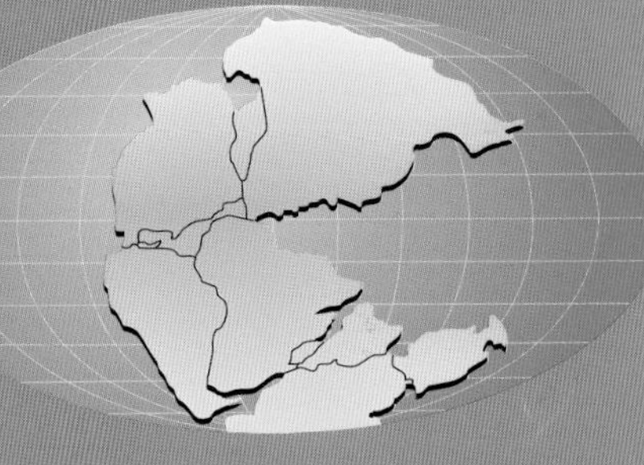

Continents 120 million years ago

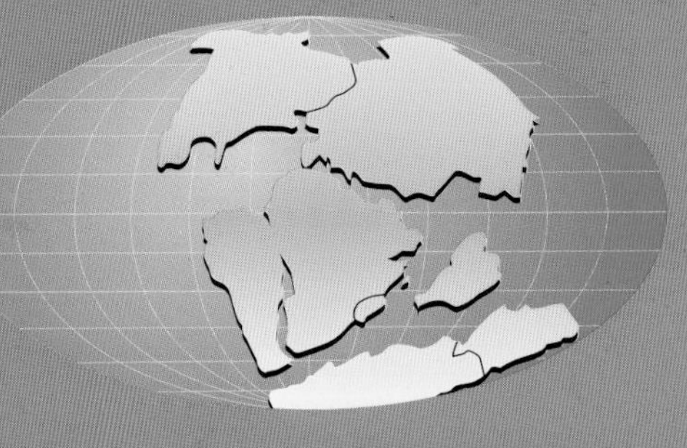

Continents today

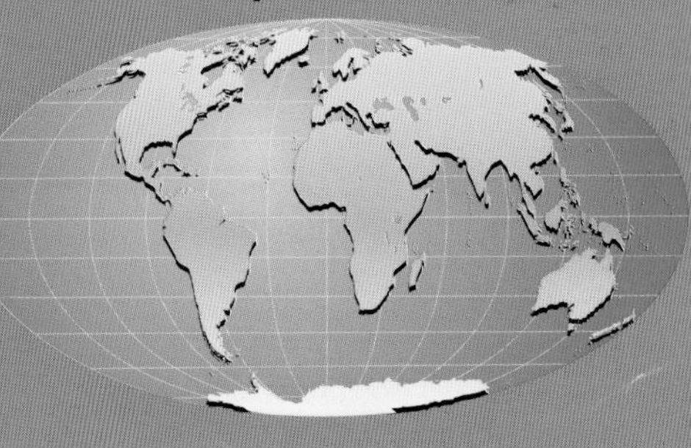

Go to Continent page 8, Inside Earth page 6, Ocean page 18

Earthquake

An earthquake happens when there are sudden movements in Earth's crust that make the ground shake. Most earthquakes are small, and we don't notice them, but during large earthquakes, rocks tremble. The land may split apart along a crack in the crust called a fault line. Some large fault lines lie above places where Earth's plates meet.

▶ A large earthquake can destroy a city and change the landscape.

3 Fires break out because **gas** pipes and electrical wires are damaged. When water pipes burst, it is difficult for firefighters to put out fires.

2 Windows break and walls crack. People find it difficult to stay standing.

1 An earthquake starts deep underground, but makes rocks on the surface crack open.

How earthquakes happen

An earthquake starts at a point called the focus. Shock waves spread out from the focus in all directions and eventually reach the surface. Rocks slip past each other and the ground splits open. The waves continue to spread and may do a lot of damage.

spreading waves

fault line

focus

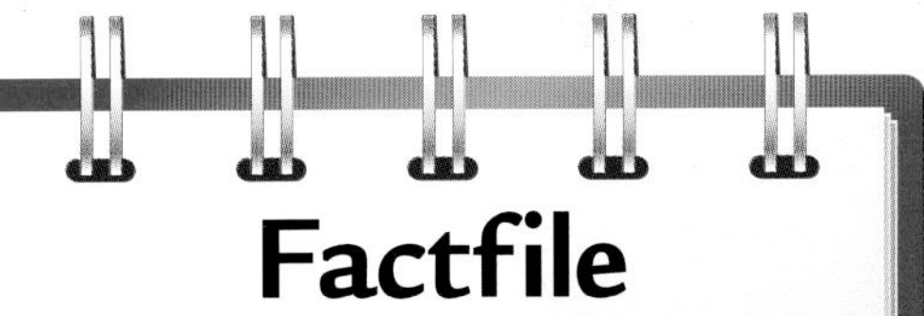

Factfile

Every year there are between 40,000 and 50,000 earthquakes on Earth, but only a few cause a lot of damage.

Most earthquakes happen less than 45 miles (72 km) underground. Some start almost as deep as 435 miles (700 km)!

Animals sometimes behave strangely before an earthquake. They may be able to sense the ground shaking before we can.

Some buildings are specially shaped to help stop them from toppling over.

4 **When the shaking becomes stronger, buildings crumble and bridges collapse.**

Quakes under the sea

Earthquakes under the ocean floor may cause huge waves called *tsunamis*. These waves travel at speeds of almost 500 miles (800 km) per hour. When a tsunami hits a sloping shore, it can build up to a height of almost 100 feet (30 m) and destroy towns on the coast.

focus

Go to Inside Earth page 6, Mountain page 30

Volcano

A volcano is a crack in Earth's crust through which gases and hot, molten rock pour out. Deep underground, in Earth's mantle, there is liquid rock, called magma. Sometimes the magma rises up through the crust and explodes into the air through a crack on the surface. This is called a volcanic eruption.

▶ **Most volcanoes erupt through cone-shaped mountains such as this.**

crater

1 Magma bursts through the top of the volcano, leaving behind a hole called a crater.

The life of a volcano

A volcano that erupts regularly is called "active." When a volcano is quiet for many years, we say it is "dormant," or "asleep." Volcanoes that are unlikely to erupt again are called "extinct."

Inside a volcano

Magma collects in an underground pool called a magma chamber. Gases in the magma force it to rise up through the crust, making chimneys called vents. Then magma explodes through the top of the volcano and becomes **lava**.

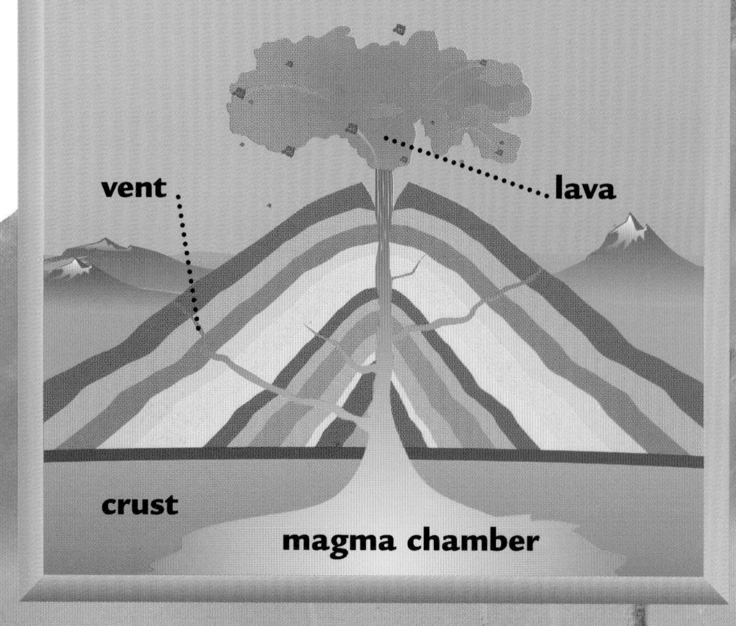

2 The volcano throws magma, ash, and steam high into the air.

Studying volcanoes

People who study volcanoes are called vulcanologists. Often, they wear special suits to protect themselves from lava. Vulcanologists record the temperature of lava and collect samples of gas and rock. This tells them more about rocks that come from inside Earth.

3 When magma reaches the surface, it is called lava. It flows down the side of the volcano.

lava

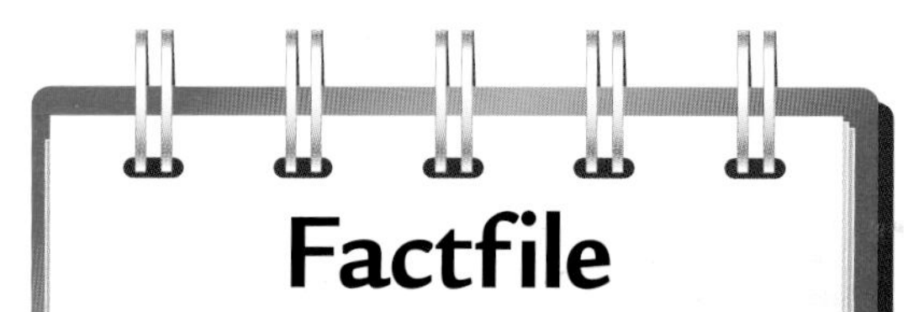

Factfile

The word "volcano" comes from Vulcan, the name of an ancient god of fire.

The highest active volcano on Earth is Ojos del Salado, on the border between Chile and Argentina. It is 22,589 feet (6,887 m) high.

Lava can be 12 times hotter than boiling water.

The Hawaiian Islands are the tops of undersea volcanoes.

4 Lava cools and hardens into volcanic rock. Over many years the rock breaks down, forming rich soil that is good for farming.

Go to Erosion page 16, Inside Earth page 6, Volcano page 12, Where on Earth? page 42

Rock

Rock is the solid part of Earth's surface and forms Earth's crust. Under the ocean, rock is covered by water, but on land rock is often covered with soil. Sometimes, in deserts and on mountains, rock is bare. Rock may be hard or soft. Many interesting things are found inside it, from rare jewels to the remains of ancient animals.

Minerals
All rocks are mixtures of tiny grains called minerals. Some minerals are soft and light, others are hard and dark. Gold and iron are types of minerals, called metals, that are found in rocks. Diamonds, and the opal shown in the photograph, are rare minerals, called gems.

Hard rocks
Marble is a rock that comes in many colors. It can be white, black, red, green, or striped. Marble is one of the world's finest kinds of rock, so it is often polished and used on the fronts of buildings. It is also a hard rock and cannot easily be worn away. For thousands of years, sculptors have carved statues from marble by carefully chiseling the rock.

Soft rocks
The Grand Canyon in Arizona is made up of many layers of rock that have been worn away by water. One of the main kinds of rock in the canyon is sandstone, which is made of sand and other pieces of rock. Sandstone is a soft rock. It can be cut and shaped easily, so it is often used for building.

Fossils

Most fossils are the remains of ancient animals and plants that have hardened in rock over millions of years and turned to stone. Sometimes, fossils are the signs left behind by animals, such as footprints, burrows, or droppings. The fossils in the photograph are the shells of ancient fish.

Digging up fossils

The word fossil comes from a word that means "to dig." Scientists dig for fossils because ancient remains can tell us what life was like on Earth millions of years ago. These scientists are chipping away the stone around fossilized dinosaur bones. The fossils will be taken to a museum where they can be carefully studied.

How fossils form

When a dead animal sinks to the bottom of a river, swamp, or ocean, the soft parts of the body rot away, leaving the bones behind. Layers of mud and sand cover the bones. As more and more layers build up, water is squeezed out. Over millions of years, the sand and mud harden and turn to stone. The bones become fossilized, which means that they turn to stone, too.

Go to Desert page 20, Mountain page 30, Ocean page 18, River page 28, Where on Earth? page 42

Erosion

Earth's surface is always changing. Oceans, rivers, ice, and the wind wear away the land by loosening rocks and breaking them up. Gradually, the rocks are broken into pieces that are small enough to be carried away. This is called erosion.

Wave erosion

As the ocean crashes against the land, waves pick up pebbles and sand and hurl them against the coast. This makes parts of the coast break off. In some places, the rock is soft and is worn away, leaving a wide curve in the land called a bay. In other places, where the rock is hard, a finger of land stands out from the coast. This is called a headland.

stack

3 When the top of the rock arch falls into the sea, a tall pillar of rock is left behind. This is called a stack.

bay

arch

headland

2 The caves grow larger and may eventually meet in the middle, leaving an arch of rock.

1 At the bottom of a headland, waves eat into the rock, making a cave. Sometimes, waves make caves on both sides of a headland.

Erosion by ice

A **glacier** is a slow-moving river of ice that usually forms high in the mountains. As a glacier slides down a mountain, the ice picks up rocks. These rocks scratch and scrape away the side of the mountain, making a deep **valley**.

snout
The end of a glacier is called the snout.

melting ice
Rocks are carried away by melting ice.

Wind erosion

In deserts, there are few plants to protect the land from the wind, so soft rock is slowly worn away. The wind also flings grains of sand at hard rocks, smoothing the edges, like sandpaper. Over years, the wind carves rocks into strange shapes, as this picture of a rock in Colorado shows.

Protecting soil

Usually, erosion happens slowly over a long period, but it can be speeded up by storms, or when trees and plants are cut down. Then, the soil may blow away, or be washed downhill. People cut steps, called terraces, into the sides of mountains, to stop soil sliding down mountainsides. In the Philippines, farmers grow rice on the terraces.

Go to Animal life page 34, Mountain page 30, Volcano page 12, Where on Earth? page 42

Ocean

The ocean is a huge area of salt water that covers most of Earth's surface. It is split into three main parts: the Pacific, Atlantic, and Indian oceans. The northern Atlantic is sometimes referred to as the Arctic Sea, and the southern part the Pacific, Atlantic, and Indian oceans is sometimes called the Antarctic or Southern ocean. The water in the oceans is always moving. At the surface, the wind whips the water into waves that crash onto the land. Deep down, warm and cold rivers of water, called currents, flow through the oceans of the world.

seamount
Under the water, there are **volcanoes** called seamounts. Waves may wear away the top of a seamount to leave a flat top.

▶ The average depth of the ocean is 12,200 feet (3,730 m). On the ocean floor, there are mountains, deep **valleys,** and huge, flat plains, just as there are on dry land.

submersible
Underwater machines, called submersibles, help people to explore life on the ocean floor.

ocean ridge
A long row of mountains runs along the middle of each ocean. This is called an ocean ridge.

▼ Life in the ocean

Seaweed floats on the ocean surface and clings to rocks on the shore. Fish eat seaweed and also use it as a hiding place.

Water near the top of the ocean is warm and sunny. Tiny animals and algae, called plankton, live here. Many sea creatures depend on plankton for food.

Mackerel and other fish live about 660 feet (200 m) below the ocean's surface. Mackerel are powerful swimmers that zoom through water looking for food.

volcanic island
When the top of a seamount breaks through the surface of the ocean, it makes an island.

continental shelf
From the edge of a **continent**, the land slopes gently downward and outward under the ocean.

trench
A trench is a long, deep valley on the ocean floor. No one knows what lives at the bottom.

continental slope
At the end of the continental shelf, the land plunges down to the deep ocean floor.

Sperm whales dive down as deep as 3,300 feet (1,000 m) to catch squid and fish. They can hold their breath for almost two hours while they search for food.

On the dark ocean floor, tripod fish wait for a meal to swim or drift past them. These fish perch on their three extra-long fins.

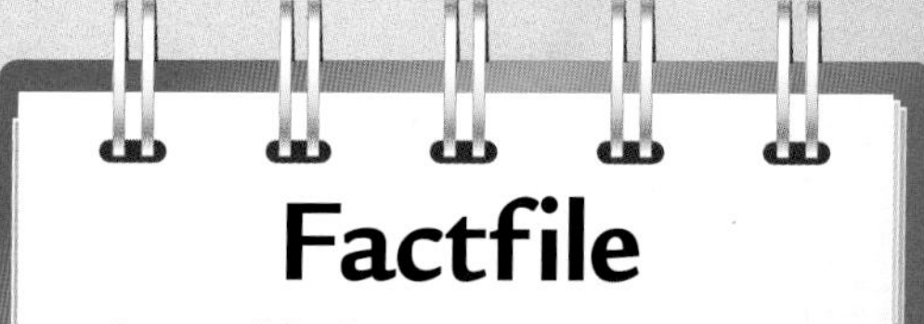

Factfile

The world's longest mountain range, the Mid-Atlantic Ridge, runs down the middle of the Atlantic Ocean.

The Pacific Ocean is bigger than all the land on Earth put together.

The deepest place on Earth is the Mariana Trench in the Pacific Ocean. It plunges down for 36,198 feet (11,033 m).

Go to Animal life page 34, Plant life page 32, Where on Earth? page 42

Desert

A desert is a dry place where it hardly ever rains. Some deserts are cold all the time, but most are hot during the day and bitterly cold at night. Many deserts are sandy, but others are covered in stones and bare rock. There is little water in deserts, so only a few plants and animals can live here. In some parts of the world, people have adapted to life in the desert.

Arizona Desert
The Arizona Desert is dry and rocky. Giant tables of hard rock stand high above the desert sands. Cactus plants thrive because they need very little water to survive. Some birds shelter from the sun by building their homes in cactus plants.

yucca
A yucca plant stores water in its stiff, narrow, pointed leaves.

cactus
A cactus stores water in its thick fleshy stem.

root system
Cactus roots spread out just below the surface and soak up rain and dew.

The Sahara

The Sahara, in Africa, is the biggest desert in the world. It covers an area almost as big as the United States. Most of the Sahara is stony, but in sandy places, the wind blows the sand into big hills called **dunes**. When the wind blows really hard, the sand flies through the air in a sandstorm.

Sand dunes

The shape of a sand dune depends on the way the wind is blowing.

1 When the wind blows from one direction, curved dunes, called barchans, form.

2 When the wind blows from all around, star-shaped dunes appear.

People in deserts

There is not enough water to grow crops in deserts, but many people live here. They move their animals and belongings from place to place in search of water and plants. These men are crossing the Syrian Desert in the Middle East. They are wearing long, loose clothes to protect themselves from the heat of the sun.

Australian desert animals

Most of the **continent** of Australia is desert. Lizards, such as the thorny devil, live here because they can survive higher temperatures than birds or **mammals**. Some animals sleep through the day, to avoid the sun. Other animals sleep through the hottest months of the year.

Go to Animal life page 34, Erosion page 16, Mountain page 30, Where on Earth? page 42

Polar lands

The polar lands are the places at the top and bottom of Earth, around the North and South poles. The land around the North Pole is called the Arctic, and the land around the South Pole is called the Antarctic. Both poles are always covered with ice and snow, making them the coldest places on Earth. Winters at the poles are long and dark, and summers are short and light.

The Arctic

The Arctic is a frozen ocean surrounded by lowland, called **tundra**, where no trees grow. Just below the surface of the tundra, the soil is always frozen. In winter, the ice in the ocean that surrounds the Arctic spreads and meets the snowy tundra. In summer, a lot of the snow and ice melts. Then, for a short time, the tundra springs to life and small, colorful plants grow.

▼ The Arctic

Key
tundra
ice

North Pole
Arctic Sea

Antarctica

Antarctica is a frozen **continent**. Most of the land is covered by a massive sheet of ice that stretches into the Southern Ocean. There are hardly any plants, but animals, such as penguins, live here and feed on fish.

▼ Antarctica

South Pole
Southern Ocean

iceberg
An iceberg is a gigantic, floating chunk of frozen fresh water. Most of an iceberg is hidden below the water's surface.

glacier
An icy **glacier** creeps along the land. When it reaches the ocean, the end breaks off and an iceberg floats away.

People at the poles
Even though the poles are the coldest places on Earth, some people live and work here. These scientists are testing the temperature and saltiness of the water.

ice floe
Jagged sheets of ice, called floes, float on top of the ocean.

ice-breaker
An ice-breaker is a special ship that cuts a path through the ice, so that other ships can follow.

bergy bits
Often, small chunks of ice, called bergy bits, break off an iceberg.

pancake ice
In places, the ice turns slushy and forms round, soft pieces that look like pancakes.

Go to Animal life page 34, Plant life page 32, Where on Earth? page 42

Grasslands

Grasslands are vast areas of flatland with few trees, where little rain falls. There are two main types of grasslands. Temperate grasslands grow in cool places. They are called prairies in the U.S., steppes in Asia, and pampas in South America. Hot, tropical grasslands are known as savannas.

giraffe
A giraffe uses its long tongue and curled upper lip to strip leaves from high branches.

African savanna
Huge herds of animals wander this grassland, nibbling the grasses and leaves. After the rains, there is plenty of grass. Some animals eat the tops of the long grasses. Other animals eat the new shoots and seeds close to the ground.

zebra
Large herds of zebra roam the savanna, feeding on the tops of the grasses.

gerenuk gazelle
Gerenuk can reach leaves and shoots on prickly bushes and low trees.

weaver birds
These birds weave blades of grass into hanging nests.

Asian steppe

Large herds of wild horses and bison used to roam the steppe in Mongolia. Today, people move from one part of the steppe to another, grazing their cattle, sheep, and goats. The people live in round tents called yurts. Some parts of the steppe have rich soil that is good for growing crops such as wheat and barley.

North American prairie

The prairie is a huge area of land. In the past, millions of bison and antelope lived here, but today the land is plowed and planted with wheat and corn to feed people. Machines are used to harvest the grain quickly.

South American pampas

The pampas does not get very hot, or very cold, which makes it the perfect place for growing corn and barley and for grazing animals. Thousands of sheep and cattle live on enormous ranches that stretch across the pampas. The cattle-farmers of Argentina are famous for rounding up the herds on horseback.

Go to

ımal life page 34, Mountain page 30, Plant life page 32, Where on Earth? page 42

Forest

A forest is a large area of land where lots of trees grow close together. Tropical rain forests grow near the equator, where it is hot and wet all year. Coniferous and deciduous forests grow in cooler places that have warm summers and cool winters. Forests are home to people and all kinds of wildlife.

Coniferous forests
Trees that grow in coniferous forests have waxy green needlelike leaves that stay on the branches all year. Conifer seeds grow in woody cones. Most of the wood that is used to build homes and to make paper comes from conifers, such as pine and spruce.

Deciduous forests
Deciduous trees, such as oak and beech, usually have wide, flat leaves. In summer, the leaves are green, but in autumn, they turn red, then golden-brown. Finally, the leaves die and fall to the ground. In winter, the branches are bare, but in spring they come back to life and new leaves grow.

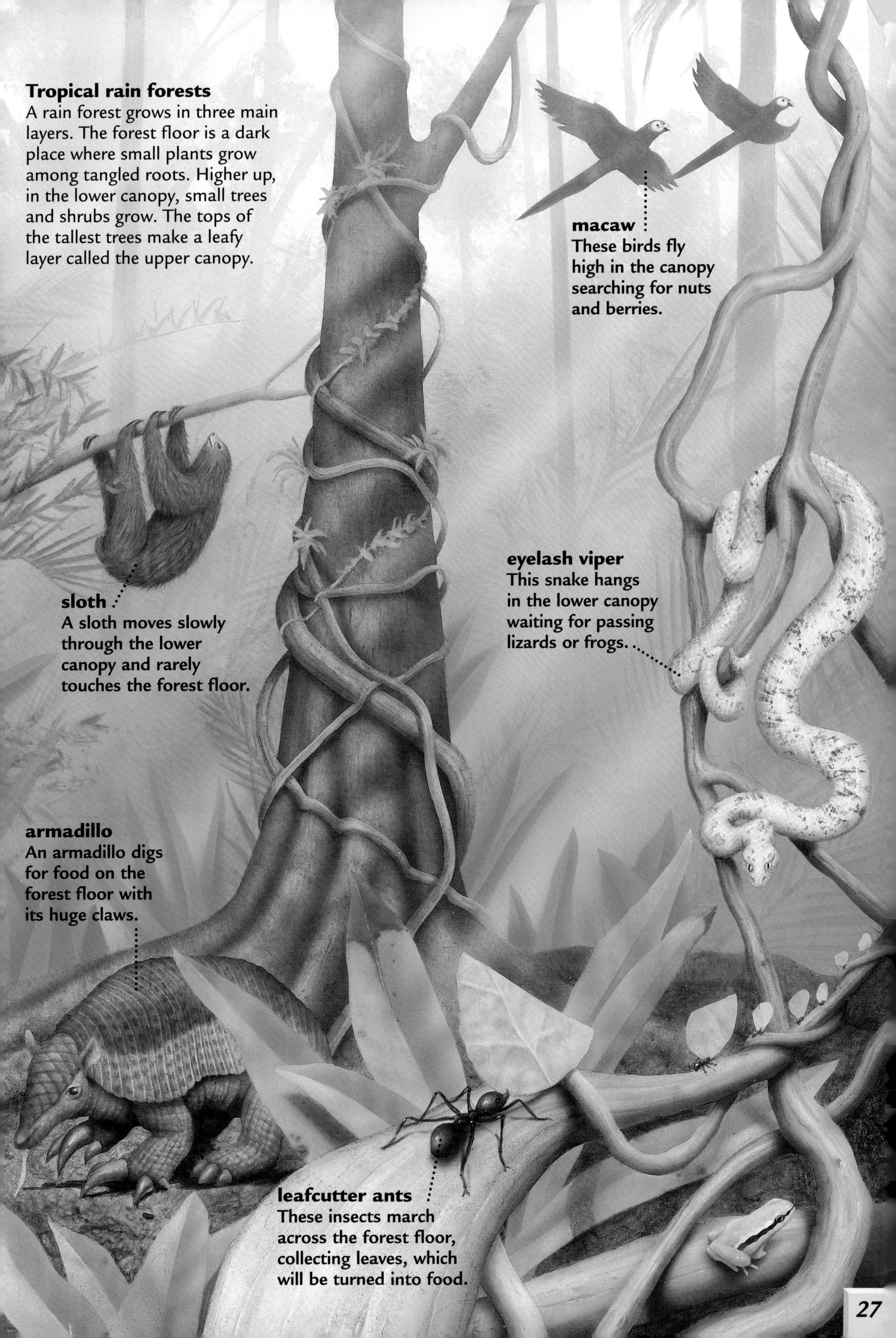

Tropical rain forests
A rain forest grows in three main layers. The forest floor is a dark place where small plants grow among tangled roots. Higher up, in the lower canopy, small trees and shrubs grow. The tops of the tallest trees make a leafy layer called the upper canopy.

macaw
These birds fly high in the canopy searching for nuts and berries.

sloth
A sloth moves slowly through the lower canopy and rarely touches the forest floor.

eyelash viper
This snake hangs in the lower canopy waiting for passing lizards or frogs.

armadillo
An armadillo digs for food on the forest floor with its huge claws.

leafcutter ants
These insects march across the forest floor, collecting leaves, which will be turned into food.

Go to Erosion page 16, Mountain page 30, Where on Earth? page 42

River

A river is a large stream of **fresh water** that flows into another river, a lake, or the ocean. The water comes from rain or snow, or even from underground. A river usually begins high in the mountains, then races downhill. As a river moves along, it changes the shape of the land, cutting into rock and breaking up the soil.

source
The start of a river is called the source. A lake is the source of this river.

valley
As a river flows down a mountain, it carves out a deep **valley** in the land.

▶ A river flows quickly through the mountains, but as more water flows into it, the river grows bigger and moves more slowly until it reaches the ocean.

waterfall
Sometimes, a river plunges over a step of hard rock, carving into soft rock below. This is called a waterfall.

meander
A river twists and turns in huge loops, called meanders, across flatland.

▼ How people use rivers

On a mountain, a fast-flowing river can be used to turn a water wheel. As the wheel turns, it moves machinery that grinds grain into flour.

A dam is a huge wall built in the mountains to hold back river water. Water flowing inside a dam can drive machines that make electricity.

Farther down the river, the water is calmer. Here, people use the river to travel from place to place, and to go fishing and sailing.

Underground rivers
Most rivers flow above-ground, but some flow underground. Certain kinds of rocks allow water to pass through them. Then, a river seeps down through the rocks and runs along underground. The river eats into rocks and carves out caves. Some caves are as big as several football fields. These are called caverns.

river mouth
The end of a river is called the mouth. When a river flows into the ocean, the fresh river water mixes with the **salt water**.

town
Often, towns grow up near rivers. Towns take their drinking water from the river.

delta
At the mouth of a river, piles of mud and sand collect in a fan shape. This is called a delta.

In places where the river is wide and slow-moving, people move heavy goods from one town, or factory, to another by boat.

Toward the end of a river, the water often floods the land, spreading rich mud over the soil. This land is good for farming and growing crops.

Factfile

The Nile River, in Africa, is 4,145 miles (6,671 km) long. It is the world's longest river.

Together, the Brahmaputra and Ganges rivers in India form the largest delta in the world.

The Colorado River has cut a valley so deep that it takes a day to walk from the top to the bottom.

Go to Continent page 8, Erosion page 16, Forest page 26, River page 28, Volcano page 12, Where on Earth? page 42

Mountain

A mountain is a part of Earth's surface that rises up steeply from the surrounding land. Mountains are much larger than hills, and take millions of years to form. Often, mountains are grouped together in long lines called chains. Some mountain chains are gradually growing taller, while others are being worn away.

mountain ridge
An icy **glacier** can tear away rock on a mountain top. When it melts, a sharp, narrow ridge is left behind.

jagged peaks
Freezing weather makes mountain rocks split to form rough, jagged peaks.

Changing mountains
As soon as mountains form, rain, wind, snow, and ice start to eat away at them. The softer rocks are worn away, leaving big dips called **valleys**. Harder rocks stand out as peaks. Young mountains usually have high, jagged tops. Old mountains are lower and more rounded because they have been worn away.

river valleys
Rivers cut down into mountain slopes, carving valleys with steep sides.

▼ Three ways mountains form

Fold mountains
When two of Earth's **plates** crash into each other, the rocks in the middle are pushed up to make fold mountains.

Volcanic mountains
When **magma** pours onto Earth's surface, it cools and hardens. Over many years, it can form **volcanic** mountains.

Block mountains
Sometimes, blocks of rock split and slide along **fault lines**. One block may be lifted higher than another to form block mountains.

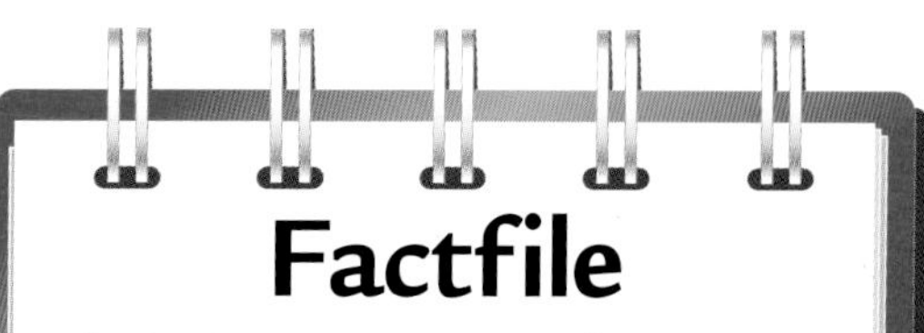

Factfile

The highest point on Earth is Mount Everest in the Himalaya of Nepal and Tibet. It is 29,028 feet (8,848 m) tall.

As you climb a mountain, the temperature of the air drops steadily.

Mount Kilimanjaro is the highest point in Africa at 19,340 feet (5,895 m) tall. It is an old volcano that is now extinct.

mountain peak
At the icy peak, it is too cold for plants to grow. There is only bare rock and snow.

meadow
Above the trees, grasses and small plants grow close to the ground, away from strong, cold winds.

upper forest
Coniferous trees grow higher up the mountainside. Their branches slope downward so that snow slides off them.

lower forest
Deciduous trees grow at the bottom of the mountain, where the air is warm.

Go to Forest page 26, Earth and its future page 40

Plant life

Plants are living things that grow all over Earth. Unlike animals, plants do not move around to find food. Instead, they are rooted in one place, and they make their own food. Most plants, from tall trees to short grasses, have flowers. The flower is the part of the plant that makes seeds, from which new plants grow. Other plants, including ferns and mosses, do not have flowers or seeds.

How a plant makes food

A plant takes a **gas** called carbon dioxide from the air and soaks up water from the soil. Leaves trap the sun's energy and use it to turn the water and carbon dioxide into the food that the plant uses to grow.

sunlight

leaves take in carbon dioxide

roots take in water

▼ There are more than 260,000 different kinds of plants, from towering trees to spiky grasses, pretty flowers, leafy ferns, and creeping mosses.

flowers
Inside each flower, there is a dust called pollen. Bees and other insects carry the pollen from one flower to another, so that new seeds can grow.

liverworts

mosses

liverworts and mosses
These small plants live in shady, moist places. They have many tiny hairlike strands, which they use to cling to surfaces.

▼ How plants scatter seeds

Blackberry seeds grow inside a fruit. Birds eat the fruit and the seeds pass out in their droppings.

Dandelion seeds float through the air like fluffy parachutes.

Seeds of the sycamore tree grow inside pods. When a pod splits open, the seeds fly far and wide.

trees
Trees are the largest and longest-living plants on Earth. Many of them grow fruit that is good to eat.

grasses
Grasses are one of the largest groups of plants. All grasses grow tiny flowers.

ferns
Ferns are one of the oldest groups of plants on Earth. They are anchored in the soil by roots that soak up water.

algae
This green film looks like a plant, but it is made of living organisms called algae.

Go to Desert page 20, Grasslands page 24, Forest page 26, Ocean page 18

Animal life

Earth is a special **planet**. The **climate**, temperature, and plant life have created the right conditions for animals to live on land, in the air, and in the ocean. Scientists have divided animals into groups to make it easy to study them. **Mammals**, fish, **reptiles**, amphibians, and birds all have backbones, and scientists call them **vertebrates**. A large group of animals, which includes insects, does not have backbones. These animals are called **invertebrates**.

Birds
Birds are the only animals that have feathers. Almost all birds can fly. When a baby bird hatches from an egg that has a hard, waterproof shell. After a few weeks, the bird's feathers develop, and its wings grow strong enough for it to take to the air. Adult birds, such as this golden eagle, spend most of their days flying.

Mammals
Mammals are the only animals that have fur or hair. Almost all mammals, such as this lioness, gives birth to live young. The mother feeds her babies on milk, which she makes in her body. There are mammals that live on land, underground, in water, and even some that can fly. People are mammals, too.

Reptiles

Crocodiles, snakes, and lizards belong to a group of animals called reptiles. Most reptiles live in hot places. Their skin is made of tough, overlapping scales that help keep their bodies moist. Some reptiles lay eggs, and others give birth to live young. Like all reptiles, a baby crocodile looks exactly like its parents, except that it is much smaller.

Amphibians

Frogs, toads, newts, and salamanders belong to a group of animals called amphibians. Like all amphibians, this frog lives both in and out of the water. On land, a frog stays in damp places because it needs to keep its skin moist. Amphibian babies hatch from eggs laid in water or on damp ground.

Insects

There are more insects on Earth than any other kind of animal. An adult insect has six legs, and its body is made up of three parts. Most insects also have two feelers, called antennae, which they use for feeling and smelling. Many insects can fly. This beetle has a hard, red wing case to protect its wings.

Fish

Fish come in many shapes, sizes and colors. They can live in near-freezing oceans or warm **tropical** seas. Like most fish, this angelfish has fins for swimming and skin made of slippery scales. A fish breathes underwater. It has openings on each side of its head, called gills, which take in a **gas**, called oxygen, from the water.

Go to Climate and season page 38, Earth and its future page 40

Weather

Sunshine, rain, wind, and snow are all different kinds of weather. The weather that happens on Earth every day is caused by the sun's rays heating the air around Earth. As the air is warmed, it moves, carrying heat and water around the world. The amount of heat and water in the air makes the weather change from place to place.

Rainbows
Rainbows happen when the sun shines through moisture in the air. The raindrops split the sunlight into all its colors. There are six main colors in a rainbow and they are always in the same order—red, orange, yellow, green, blue, and violet.

The water cycle
There is water everywhere on Earth—in rivers, in oceans, and even in the air. Water moves from the oceans and land, to the air, to the land and back to the oceans again. This is called the **water cycle**.

Types of clouds

There are three main cloud shapes that can help you tell what kind of weather may happen.

Cirrus clouds
Thin and wispy clouds show that the weather is changing.

Cumulus clouds
Fluffy clouds that look like cotton mean good weather.

Stratus clouds
Long, flat clouds low in the sky show that rain is on the way.

Wind

The wind may blow gently as a breeze, or so hard that it pushes over trees. Wind is moving air. It happens because the sun heats the land, which warms the air above it. In hot places, warm air rises. Then, cool air from colder places rushes in to take the place of the warm air. We feel this moving air as wind.

Electrical storms

Bad weather can bring storms with black clouds, strong winds, heavy rain, and even thunder and lightning. A flash of lightning is a huge spark of electricity in the sky. Lightning heats up the air around it quickly, shaking the air and making the sound we call thunder.

Go to Animal life page 34, Weather page 36

Climate and season

Climate is a pattern of **weather** that is roughly the same from one year to the next. Some places have warm climates, and others have cold climates. The climate of a particular place depends mostly on where it is in the world. In many places, the weather changes throughout the year. These changes are called **seasons.**

World climates
Places near the **equator** are hot and wet all year. They have a **tropical** climate. At the poles, the climate is cold and dry. In between, the climate is usually **temperate**, which means that it is mild and damp.

Key

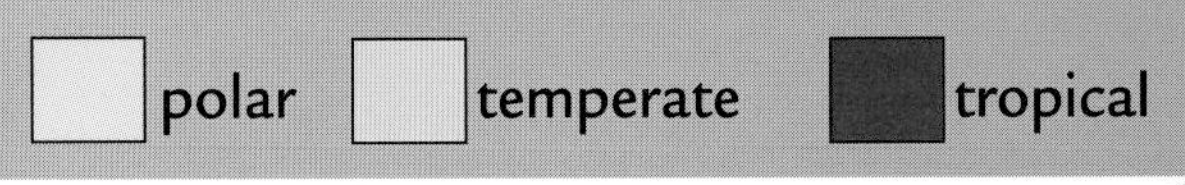

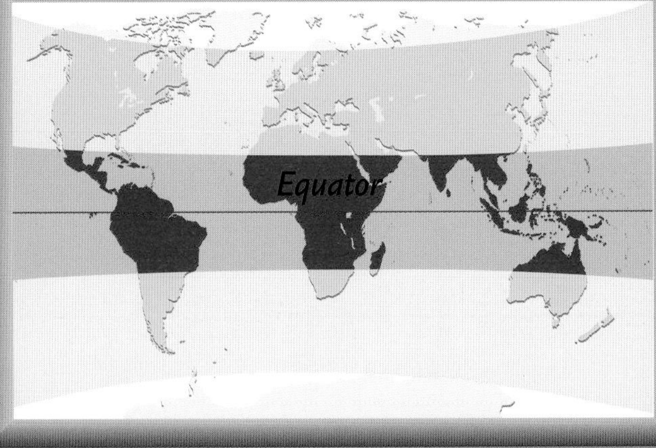

▼ This picture shows why some places on Earth have cold climates, and others have hot climates.

The sun warms Earth all over, but shines more strongly on some parts of it than on others. This is because Earth is curved.

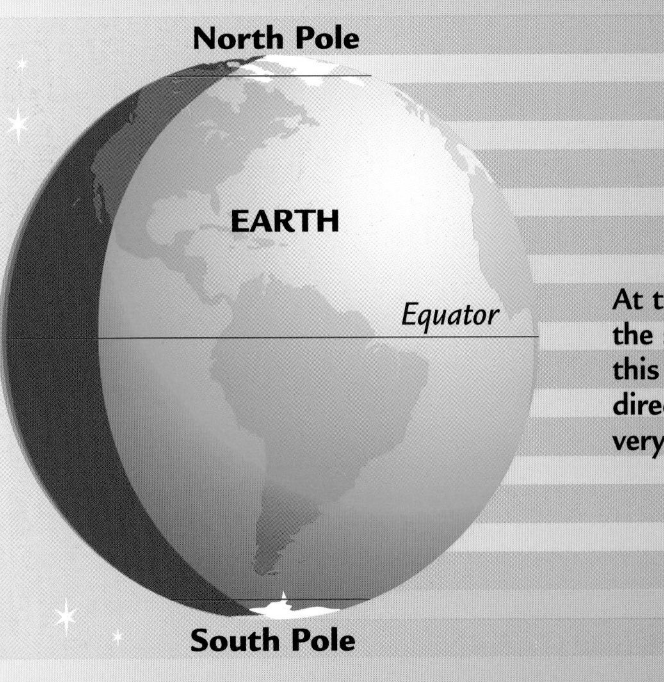

At the equator, the sun's rays hit this part of Earth directly, making it very hot.

SUN

At the North and South poles, the sun's rays are spread out, and give less heat to this part of Earth. Polar lands are very cold.

Four seasons
Temperate places have four seasons. In winter, days are short and cold. Spring days are longer and brighter. Summer days are warm and sunny. Cool weather returns in autumn.

Spring
On a cherry tree, small leaves and buds appear.

Summer
The tree keeps its leaves over the summer season.

Autumn
Leaves turn golden, then brown, die and drop off.

Winter
The bare tree stores its energy until spring.

Hibernation
Some animals change their habits with the seasons. During cold winter months, chipmunks, hamsters, and hedgehogs fall into a deep sleep called hibernation. They live off fat stored in their bodies. Hibernating animals wake up when spring arrives and days are warmer.

Wet and dry seasons
Some hot places near the equator, such as India, have two main seasons in a year, one dry and one wet. These are called monsoon seasons. In the winter monsoon, hot, dry winds bake the land. During the summer monsoon, fierce winds bring clouds and lots of rain. Often there are floods in towns and cities.

Go to Forest page 26, Plant life page 32, Weather page 36

Earth and its future

Many millions of people live on Earth, and the way we live is changing the planet. People are cutting down forests and allowing poisonous gases to pollute Earth's atmosphere. We need to repair the damage and look after the world's natural resources for the future.

Acid rain

In some places, gases from cars and factories mix with water in the air. The gases turn to acid and become part of the **water cycle**, falling as **acid rain**. This can kill wildlife and destroy whole forests.

Global warming

There is a layer of gas around Earth, which traps some of Earth's heat. Fumes from cars, factories, and burning forests mix with the gas layer and trap even more heat. Slowly, this is making Earth warm up. This is called global warming. If the water in the oceans warms up too, then the level of the sea will rise, and towns and cities on the coast may be flooded.

some heat escapes from Earth

layer of gases

gases from factories, cars, and burning forests rise and mix with the air

heat bounces back to Earth

Ozone holes

High above Earth, there is a layer of gas called **ozone**. Ozone works like a shield, protecting people, animals, and plants from the harmful rays of the sun. Gases, called CFCs, from factories, refrigerators, and aerosols destroy ozone. This picture was taken from space. The dark blue circle above Antarctica shows a growing hole in the ozone layer.

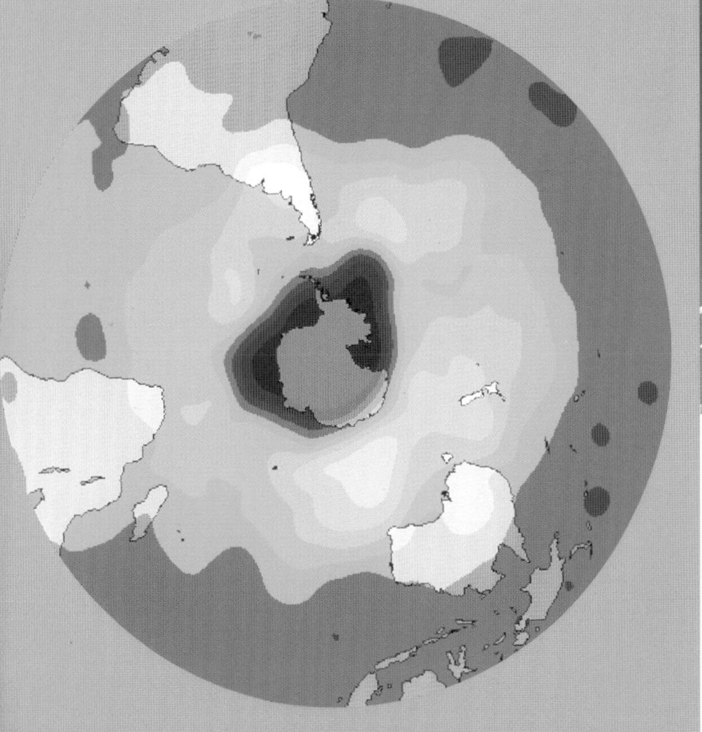

Wind power

We use **oil** and gas, which are natural resources, to make electricity. These fuels pollute the atmosphere, and one day they may also run out. Today, many countries make electricity using energy from the sun, water, and wind. This energy is much cleaner and will not run out. These windmills use the wind to make enough electricity to supply thousands of homes.

Planting trees

Many people are working hard to protect Earth and to look after its resources. The boy in this picture is growing trees from seeds to plant in the forest. Trees are important because they give off a gas called oxygen, which all living things need to survive. Too many trees have been cut down in the past for wood, or to make space for farmland. Now, many more trees need to be planted for the future.

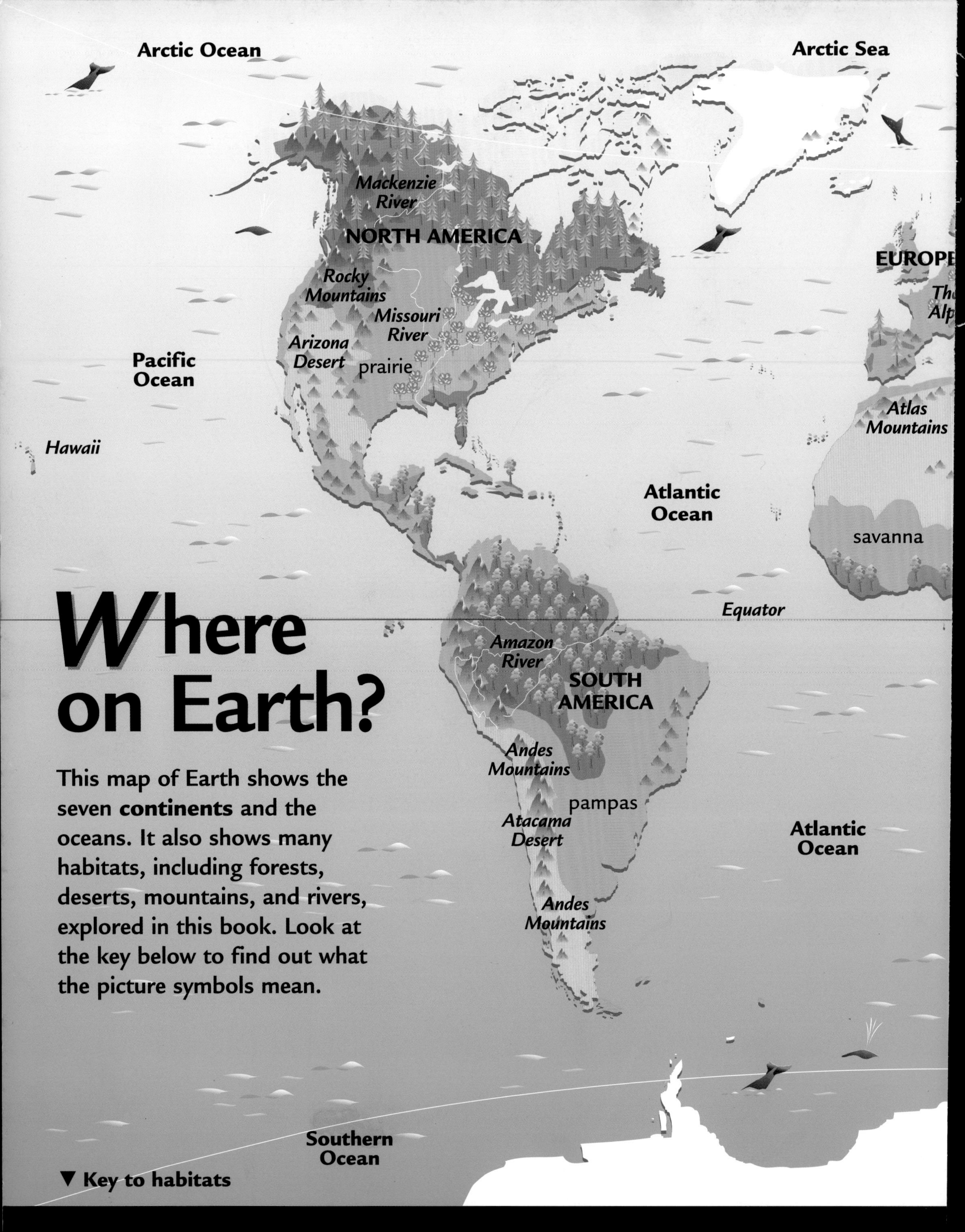

Where on Earth?

This map of Earth shows the seven **continents** and the oceans. It also shows many habitats, including forests, deserts, mountains, and rivers, explored in this book. Look at the key below to find out what the picture symbols mean.

▼ **Key to habitats**

A river is a stream of **fresh water** flowing into a lake, another river, or the ocean.

Tundra is land in the northern part of the world that is frozen underground.

A desert is a place where it rarely rains. Here, it is hot in the day and cold at night.

Grassland is a huge, flat area of land called savanna, pampas, prairie, or steppe.

A rain forest is a hot, wet forest that grows at, or near, the **equator**.

A deciduous forest is a forest of trees that lose their leaves in autumn and grow new ones in spring.

A coniferous forest is a forest of trees that keep their leaves and stay green all year.

A mountain chain is a long line of mountains that are grouped together, roughly side by side.

Polar lands are icy places at the top and bottom of the world, around the North and South poles.

Amazing facts

On these pages, you will discover amazing facts about the history of the Earth and about some of the planet's most unusual features. You will learn how many people live on Earth, and what may happen to it in the future.

Journey to the Earth's center

Earth's center is 4,000 miles (6,400 km) away. Starting at the crust, if you walked non-stop, you would reach the Earth's center in about 128 days. It would take you...

EARTH

...half a day to walk through the crust.

...58 days to walk through the mantle.

...70 days to walk through the core to the center.

center

crust

core

mantle

Earth File

Highest waterfall
Name: Angel Falls
Place: Venezuela, South America
Fact: The water drops a total of 3,212 feet (979 m).

Saltiest Sea
Name: Dead Sea
Place: Southwest Asia
Fact: This sea is so salty that people cannot sink in it.

Tallest sand dunes
Name: Isaouane-N-Tifernine
Place: Algeria, North Africa
Fact: The highest dunes reach 1,526 feet (465 m).

Fastest moving glacier
Name: Quarauac Glacier
Place: Greenland
Fact: The glacier slides downhill, up to 79 miles (24 m) in a day.

Most active volcano
Name: Stromboli
Place: Mediterranean Sea
Fact: When the volcano is active, it erupts continually for months, or even years at a time.

Earth timeline

Earth formed about 4.5 **billion** years ago, but this is so long ago that it is difficult to imagine. It is easier to think of short lengths of time, such as hours or days. Imagine that Earth began 24 hours ago, at midnight. These clocks tell you when everything else formed.

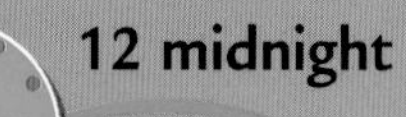

At 12:00 midnight, Earth began. At first it was a fiery, hot ball.

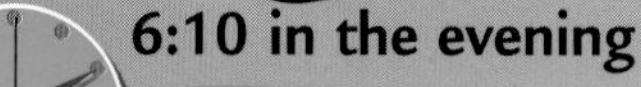

At 6:10 p.m., over 18 hours later, jellyfish and other simple animals appeared.

9:50 at night

At 9:50 p.m., the first fish, including sharks, swam in the oceans.

How many people?

Fifty years ago, about 2.5 billion people lived on the planet. Better medicines and more food meant that many people began to live longer.

Today, there are almost 6 billion people on Earth—more than double the number of 50 years ago. Many people live longer than ever before.

If the number of people on Earth continues to grow at this rate, then 100 years from now, there will be an incredible 12 billion people on Earth.

50 years ago

today

100 years from now

Strange weather

When storms rage and the weather is dramatic, strange things can happen.

Terrible twister

In 1931, in the U.S., a tornado lifted a train that weighed 91 tons (83 metric tons) into the air and dropped it in a ditch.

Heavy hailstones

The heaviest hailstones ever recorded fell in Bangladesh in 1986. They weighed over 2 pounds (1 kg), about as much as a small melon.

Flying frogs

A shower of frogs fell from the sky in England in 1954. Strong winds picked up the frogs from their watery homes and carried them through the air.

What will happen to Earth in the future?

The world is slowly warming up. If Earth's temperature rises by 7 °F (4 °C), this could heat up the water in the oceans and make it rise. Cities such as New York could be flooded.

Thousands of years from now, Earth may cool down. There could be a big ice age, which means that ice would cover much of Earth's surface. The last ice age was 10,000 years ago.

10:15 at night

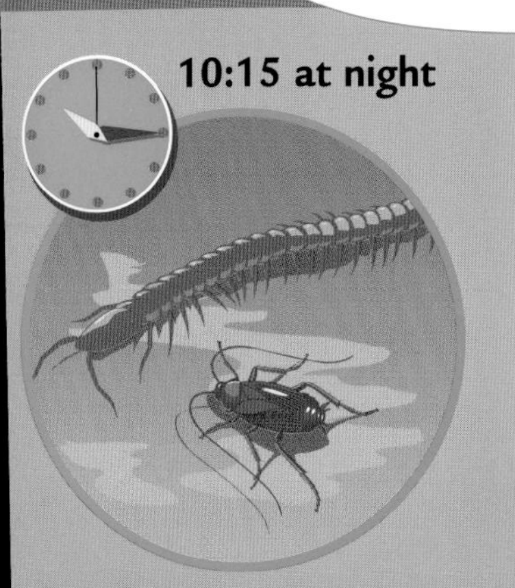

At 10:15 p.m. giant insects and millipedes up to 2m long crawled across Earth.

10:30 at night

At 10:30 p.m. the first trees started to grow. They were **coniferous** trees.

10:45 at night

At 10:45 p.m. the first dinosaurs roamed the planet.

11.59.55 at night

At 5 seconds to midnight, or just 5 seconds ago, people like us appeared.

Index

Acid rain 40, 46
Africa 8, 24, 29, 31, 44
African Plate 8
air 4, 5, 31, 32, 33, 36, 37, 40, 43
algae 33
Algeria 44
Alps 42
Amazon River 42
amphibian 34, 35
Andes Mountains 42
Angel Falls 44
angelfish 35
animal 4, 5, 7, 11, 14, 15, 18, 19, 20, 21, 22, 24, 25, 27, 32, 34–35, 39
Antarctica 8, 22, 41, 43
Arabian Desert 43
arch 16
Arctic Sea 18, 22, 42, 43
Argentina 13, 25
Arizona Desert 20, 42
armadillo 27
Asia 8, 24, 25, 43
Atacama Desert 42
Atlantic Ocean 18, 19, 42
Atlas Mountains 42
atmosphere 40, 46
Australia 8, 21, 43
Australian Desert 43

Bacteria 4
bald eagle 34
Bangladesh 45
Barren Desert 43
bay 16
bedrock 7
bee 32
beetle 35
bergy bits 23
big bang 4, 46
bird 5, 20, 21, 24, 27, 33, 34
bison 25
block mountain 30
Brahmaputra River 29

Cactus 20
canopy 27
canyon 14
carbon dioxide 32
cave 16, 29
cavern 29
CFC 41
Chile 13
chipmunk 39
city 10, 39, 40, 45
cliff 16
climate 34, 38–39, 46
cloud 4, 5, 36, 37, 39
coast 11, 16, 40
Colorado 17
Colorado River 29
Congo River 43
coniferous forest 26, 31, 43, 45
continent 8–9, 19, 22, 42, 46
continental shelf 19
continental slope 19
crater 12
crocodile 5, 35
crop 21, 25, 29
crust 5, 6, 7, 8, 9, 10, 12, 14, 44, 46
current 18, 46

Dam 28
dandelion 33
Danube River 43
Dead Sea 44
deciduous forest 26, 31, 43
delta 29
desert 14, 17, 20–21, 42
diamond 14
dinosaur 5, 15, 45
dune 21, 44, 46

Earthquake 9, 10–11
electricity 7, 28, 37, 41
England 45
Equator 26, 38, 39, 42, 43, 46
erosion 16–17, 46
eruption 12
Ethiopian Highlands 43
Europe 8, 42
eyelash viper 27

Factory 29, 40, 41
farming 13, 17, 25, 29, 41
fault line 9, 10, 30, 46
fern 32, 33
fire 10
firefighter 10
fish 4, 15, 18, 19, 22, 34, 35, 44
flood 39, 40
flower 32, 33
fold mountain 30
forest 26–27, 31, 40, 41, 42, 43
fossil 14–15
fresh water 22, 28, 29, 42, 46
frog 27, 35, 45
fuel 7, 41

Ganges River 29, 43
gas 4, 5, 7, 12, 13, 32, 35, 36, 40, 41
gem 14
gerenuk gazelle 24
giraffe 24
glacier 17, 23, 30, 44, 46
global warming 40
goat 25
Gobi Desert 43
gold 14
Grand Canyon 14
grain 28
grass 24, 31, 32, 33
grassland 24–25, 42
Greenland 44

Hailstone 45
hamster 3
Hawaiian Islands 13
headland 16
hedgehog 39
hibernation 39
Himalaya 8, 31, 43
human 5, 6, 10, 13, 20, 21, 23, 25, 26, 28, 29, 34, 40, 41, 45

Ice 4, 16, 17, 22, 23, 30
ice age 45
ice floe 23
iceberg 22, 23
ice-breaker 23
Iceland 9
India 29, 39
Indian Ocean 18, 43
inner core 6
insect 32, 34, 35, 45
invertebrate 34, 46
iron 14
Isaouane-N-Tifernine 44
island 13, 19

Jellyfish 4, 44
jewel 14

Kalahari Desert 43
Kara Kum Desert 43

Lake 28, 42
land 4, 6, 8, 9, 10, 14, 16, 17, 19, 22, 23, 24, 25, 26, 28, 29, 34, 35, 39
lava 7, 12, 13, 46
leaf 20, 24, 26, 32, 39, 43
leafcutter ant 27
lightning 37
lion 34
liverwort 32
lizard 21, 27, 35

Macaw 27
machine 7, 18, 25, 28
Mackenzie River 42
mackerel 18
magma 7, 12, 13, 30, 46
mammal 5, 21, 34, 46
mantle 6, 7, 8, 12, 44, 46
marble 14
Mariana Trench 19
meander 28
Mediterranean Sea 44
metal 6, 7, 14
Mid-Atlantic Ridge 19
Middle East 21
mineral 14
mining 7
Missouri River 42
Mongolia 25
monsoon 39
moss 32
Mount Everest 31
Mount Kilimanjaro 31
mountain 8, 12, 14, 17, 18, 28, 30–31, 42, 43
mountain chain 30, 43
mountain peak 8, 30, 31
mountain ridge 30
mountain terrace 17
mouth *of a river* 29
Murray-Darling River 43

Namib Desert 43
natural resources 40, 41, 46
Nepal 31
New York 45
newt 35
Nile River 29, 43
North America 8, 20, 25, 42
North Pole 22, 38, 43

Ob River 43
ocean 4, 5, 6, 8, 9, 11, 14, 15, 16, 18–19, 22, 23, 28, 29, 34, 35, 36, 40, 42, 43, 44, 45
ocean floor 11, 18, 19
ocean ridge 18
oil 7, 41, 46
Ojos del Salado 13
opal 14
outer core 6
oxygen 35, 41
ozone 41, 46

Pacific Ocean 18, 19, 42, 43
pampas 24, 25, 42
pancake ice 23
Pangaea 9
paper 26
penguin 22
Philippines 17
plain 18
planet 4, 6, 34, 40, 45, 46
plankton 18
plant 4, 7, 15, 17, 18, 20, 21, 22, 26, 27, 31, 32–33, 34, 41
plate *of the Earth* 8, 9, 10, 30, 46
polar lands 22–23, 43
pollen 32
pollution 40, 41
power station 7
prairie 24, 25, 42

Quarauac Glacier 44
quarry 7

Rain 5, 20, 24, 28, 30, 36, 37, 39, 40
rainbow 36
rain forest 26, 27, 42
reptile 5, 34, 35, 46
rice 17
river 16, 28–29, 30, 42
rock 4, 5, 6, 7, 10, 12, 13, 14–15, 16, 17, 18, 20, 28, 29, 30, 31
Rocky Mountains 42
root 7, 20, 27, 32, 33

Sahara Desert 21, 43
salamander 35
salt water 18, 29, 46
San Andreas Fault 9
sand 14, 15, 17, 20, 21, 29
sand dune *see* dune
sandstone 14
sandstorm 21
savanna 24, 42
scientist 4, 9, 15, 23, 34
sea *see* ocean
seamount 18
season 38–39, 46
seed 24, 26, 32, 33, 41
shark 44
sheep 25
shell 15
shore 11, 18
sloth 27
snake 27, 35
snout *of a glacier* 17
snow 22, 28, 30, 31
soil 6, 7, 13, 14, 17, 22, 25, 28, 29, 32, 33
source *of a river* 28
South America 8, 24, 25, 42, 44
South Pole 22, 38, 43
Southern Ocean 18, 22, 42, 43
sperm whale 19
stack 16
steppe 24, 25, 43
storm 17, 37, 45
Stromboli 44
submersible 18
Sun 4, 20, 21, 32, 36, 37, 38, 41
supercontinent 9
sycamore tree 33
Syrian Desert 21, 43

Temperate climate 38, 39
temperate grasslands 24
thorny devil 21
thunder 37
Tibet 31
Tien Shan Mountains 43
tornado 45
town 11, 29, 39
tree 17, 22, 24, 26, 27, 31, 32, 33, 39, 41, 43, 45
trench 19
tripod fish 19
tropical climate 38
tropical grasslands 24
tropical rain forest *see* rain forest
tsunami 11
tundra 22, 42, 46
turtle 5

United States 9, 14, 17, 21, 24, 29, 45

Valley 17, 18, 19, 28, 30, 46
Venezuela 44
vertebrate 34, 46
volcanic island 19
volcanic mountain 30
volcanic rock 13
volcano 7, 12–13, 18, 31, 44, 46
Vulcan 13
vulcanologist 13

Water 5, 13, 14, 15, 18, 20, 21, 22, 23, 28, 29, 32, 33, 34, 35, 36, 40
water cycle 36, 40, 46
water vapor 36
water wheel 28
waterfall 28, 44
wave 10, 11, 16, 18
weather 36–37, 38, 39, 45, 46
weaver bird 24
wind 16, 17, 18, 21, 30, 31, 36, 37, 39, 41, 45
wind power 41
windmill 41
wood 26, 41

Yangtze River 43
Yellow River 43
yucca 20
yurt 25

Zebra 24

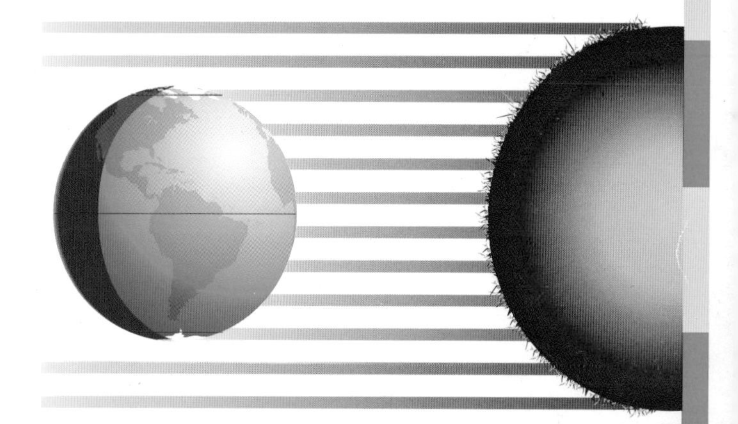

Troubleshooting tips

System requirements

The Interfact Reference Earth disk will run on most Windows or Apple Macintosh computers. To check that it will run on yours, please read the minimum specifications below.

Windows

486DX2/66Mhz PC with Windows version 3.1, 3.11, 95 or 98; VGA color monitor; SoundBlaster-compatible soundcard; 8Mb RAM (16Mb RAM recommended with Windows 95; 24 Mb recommended with Windows 98).

Macintosh

Apple Macintosh with 68020 processor (or Power Macintosh), system 7.0 (or later) and 16Mb of RAM (24 Mb with Power Macintosh).

Settings

To get the most out of your Earth disk, please check:

Your monitor is set to 640x480 and 256 colors.

You have the Arial font (Windows users) installed in your fonts folder.

You have no other applications open.

Common problems

Symptom: Graphics freeze or text boxes appear blank (Windows 95 or 98 only).
Problem: Graphics card acceleration too high.
Solution: Right-click on "My Computer." Click "Properties," then "Performance" then "Graphics." Reset the hardware acceleration slider to "None." Click OK. Then restart your computer.

Symptom: There is no sound (Windows users only).
Problem: Your sound card is not SoundBlaster-compatible.
Solution: Try to configure your sound settings to make them SoundBlaster-compatible (refer to your sound card manual for more details).

Read me file

If you have a problem with the Earth disk that is not covered in the notes above, be sure to check the Read me file. On Macintosh computers and PCs that do not start the Earth disk automatically, you will find the Read me file next to the Earth icon.

To access the Read me file on PCs that start the Earth disk automatically, hold down the shift key when you insert the disk. Click on My Computer, then right-click on the CD icon. Click on Open to see the contents of the CD, then click on the Read me icon.

Helpline

If you come across a problem loading or running the Earth disk, you should find the solution here. If you still cannot solve your problem, call the helpline at 1-609-921-6700. Remember to get permission from the person who pays the bill before you use the phone.

Published in the United States and Canada by
Two-Can Publishing LLC
234 Nassau Street
Princeton, NJ 08542

www.two-canpublishing.com

For information on Two-Can books and multimedia, call 1-609-921-6700, fax 1-609-921-3349, or visit our web site at http://www.two-canpublishing.com

Created by
act-two
346 Old Street
London EC1V 9RB

Disk
Multimedia Director: William Wharfe
Art Director: Sarah Evans
Senior Project Editor: Rob Mitchall
Senior Designer: James Evans
Illustrators: Michael Egar, James Evans, Richard Harris, Carlo Tartaglia
Programmers: Roger Emery, Kirsten Minshall
Consultant: Tim Davy
Production Director: Lorraine Estelle
Production Manager: KatherineHarvey

Book
Art Director: Belinda Webster
Managing Editor: Deborah Kespert
Senior Commissioning Editor: Jacqueline McCann
Senior Designer: Helen McDonagh
Designer: Michele Egar
Editorial Assistant: Lucy Arnold
Picture Research: Laura Cartwright
Consultant: Carmela Di Landro
Main illustrations: Brian McIntyre
Computer illustrations: Mel Pickering, Jacqueline Land
Large computer illustrations: Sebastian Quigley

Two-Can Publishing is a division of Zenith Entertainment plc, 43-45 Dorset Street, London W1H 4AB

ISBN 1-58728-472-3

1 2 3 4 5 6 7 8 9 10 03 02 01 00

Photographic credits:
Bruce Coleman Ltd/Jeff Foot Productions p14b, BCL/Keith Gunner p15br, BCL/John Shaw p26, BCL/Michael Fogden p35t, BCL/Andrew Purcell p35b, BCL p37t, BCL/George McCarthy p39t; NHPA p14tl, NHPA/Gerard Lacz p34b; Oxford Scientific Films/Warren Faidley p37b; Pictor International p21b; Planet Earth Pictures/I & V Krafft/Hoa Qui p7t, PEP p13, PEP/Adam Jones p17t; Robert Harding Picture Library/Michio Hoshino p34t, RHPL/JHC Wilson p39b; Science Photo Library/Simon Fraser p9t, p40, SPL/Sinclair Stammers p15t, SPL/NOAA p41tl, SPL/Marcelo Brodsky p41b; Telegraph Colour Library p41tr; The National Geographic Society/David S Boyer p29; The Stock Market p7b; Tony Stone Images/Chris Noble p8, TSI/James Balog p9b, TSI/AB Wadham p14tr, TSI/John Callahan p17b, TSI/Greg Probst p22, TSI p23, TSI/James Strachan p25t, TSI/Andy Sacks p25b, TSI/Gary Yeowell p36; Zefa/Spichtinger p21t.

Printed in Hong Kong